目　次

1、ペット用の紙おむつ「腹当て部型」の使用方法 ────────────── 3

　⑴　ペットの腹部にあてがう ─────────────────────── 3

　⑵　ペットから排出された尿〜 ────────────────────── 3

　⑶　ペットの尻尾の部分の隙間〜 ───────────────────── 4

　⑷　ペットから排出された糞〜 ────────────────────── 4

　⑸　おむつの取り外しのケア〜 ────────────────────── 5

2、ペット用の紙おむつ「パンツ型」の使用方法 ───────────────── 7

　⑴　腹当て部を下向き〜 ──────────────────────── 7

　⑵　尻尾の部分は〜 ─────────────────────────── 7

　⑶　腹当て部の弛む部分〜 ─────────────────────── 8

　⑷　尻尾の部分の隙間〜 ──────────────────────── 8

　⑸　ペットの後ろ足２本〜 ─────────────────────── 9

3、使用方法の英文 ─────────────────────────────── 11

　Usage of the paper diaper "abdomen type" for pets ───────────── 11

　Usage of the paper diaper "underwear type" for pets ──────────── 15

4、公報解説　ペット用の紙おむつ・特許第４２０１７９８号

　⑴　請求の範囲 ───────────────────────────── 17

　⑵　発明の詳細な説明 ───────────────────────── 18

　⑶　技術分野 ────────────────────────────── 18

　⑷　背景技術 ────────────────────────────── 19

　⑸　発明が解決しようとする課題 ─────────────────── 19

- (6) 発明が解決するための手段 ････････････････････････････････19
- (7) 発明の効果 ･･21
- (8) 発明を実施するための最良の形態 ････････････････････････21
- (9) 図面の簡単な説明 ･････････････････････････････････････26
- (10) 符号の説明 ･･･26

5、英文 ･･35

1、ペット用の紙おむつ「腹当て部型」の使用方法

(1) ペットの腹に腹当て部をあてがいう。 この場合、尻尾が尻当て部の中央の割れ目から外に出るようにし、尻当て部を尻から背中に回して、尻当て部の左右の縁を腹当て部に左右の縁に粘着テープで止める。

(2) ペットから排出された尿は、腹当て部の内側に取り付けられている吸水性樹脂により吸収される。

(3) ペットの尻尾の部分の隙間による排泄物の漏れ防止には、尻尾の隙間の周縁はシャーリングにより塞がれるので漏れを防げる。

(4) ペットから排出された糞は、腹当て部の内側に取り付けられている吸水性樹脂の部分に落下し吸収されるので、ペットの皮膚への汚れが少なく、おむつのケアが容易にできる。

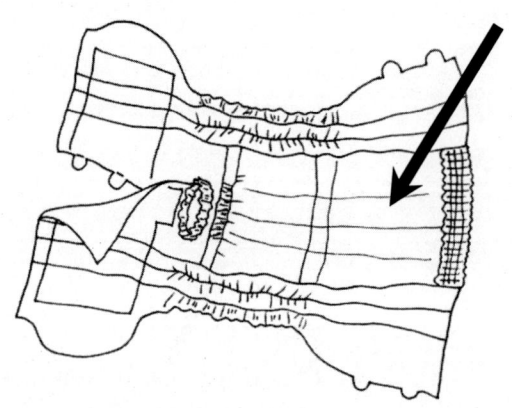

⑸ おむつの取り外しのケアは、「１」の逆順に行いますが、接着した部分を剥がして処理する。ペットの皮膚は濡れテッシュで拭き取り清潔にする。

２、ペット用の紙おむつ「パンツ型」の使用方法

(1) 腹当て部を下向きにし、ペットの足から順番にパンツを穿かせる。

(2) 尻尾の部分は、パンツの尻当て部の中央孔から尻尾を外側に引き出して装着する。

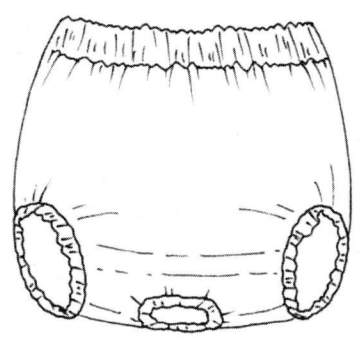

⑶　腹当て部の弛む部分はゴム帯で伸縮するようになっている。

⑷　ペットの尻尾の部分の隙間による排泄物の漏れ防止には、尻尾の隙間の周縁はシャーリングにより塞がれるので漏れを防げる。

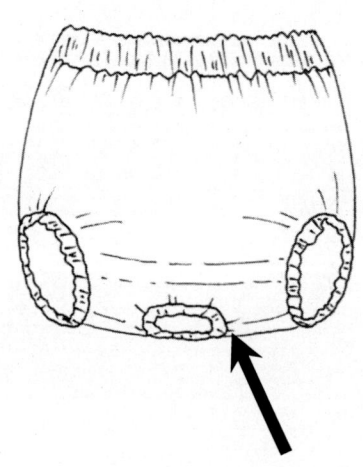

⑸　ペットの後ろ足2本を通すための開口部の隙間は、排泄物の漏れを防止できるよう、シャーリングになっている。

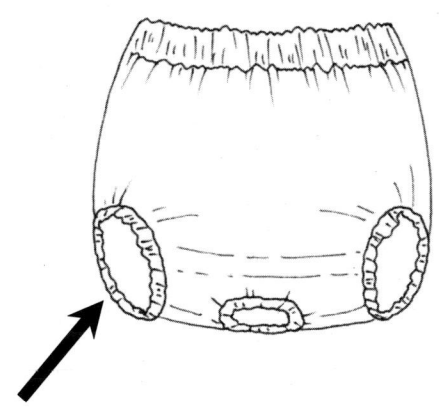

Usage of the paper diaper "abdomen type" for pets

(1)

put it on in the stomach of the pet .

A tail goes out of the central crack of the buttocks sheet and invests it in a back from buttocks and stops the relationship of relationship of right and left of the buttocks cloth and right and left of the stomach cloth with adhesive tape.

(2)

The urine drained from a pet is processed by an absorbing water sheet got inside of the abdomen.

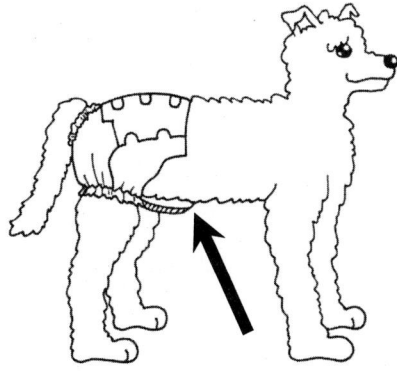

(3)

Because it is blocked up by shirring, for prevention of leak of the excrement by the gap of the part of the tail of the pet, the fringe of the gap of the tail can prevent a leak.

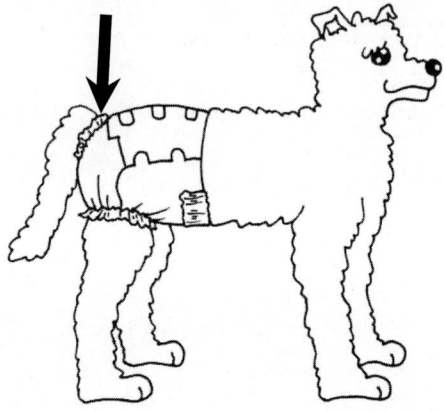

(4)

Because the feces drained from a pet fall to water absorptive resin attached inside of abdominal cloth and are absorbed, there are few dirts to skin and can facilitate the care of the diaper.

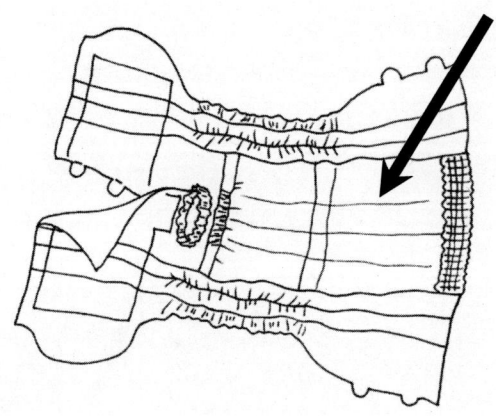

(5)

I perform the care of the disassembly of the diaper in "1" inverse order, but I tear off the part which adhered and handle it. I wipe off the skin of the pet with a wet tissue and keep it clean.

Usage of the paper diaper "underwear type" for pets

(1)

I lower abdominal cloth and let you wear underwear from the foot of the pet in turn.

(2)

The part of the tail drags a tail out of the central aperture of the lining added to the seat of garments for reinforcement part of the underwear outward and puts it on.

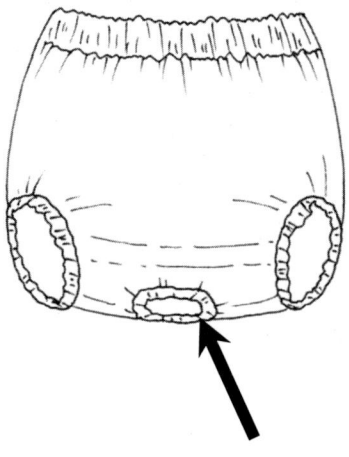

(5)

The gap of the opening to maintain two hind legs of the pet becomes the shirring to prevent a leak of the excrement.

2 and 3 are omitted

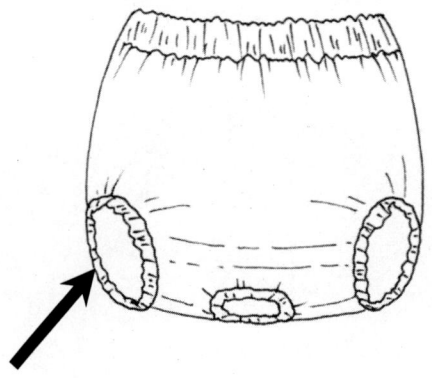

3、公報解説　ペット用の紙おむつ・特許第４２０１７９８号

特許第４２０１７９８号

発明の名称；ペット用の紙おむつ

特許権者／発明者；橋本　みち子

⑴【特許請求の範囲】

【請求項１】

四足ペットの腹から後足の股にあてがう腹当て部と、尻にあてがう尻当て部とが一体状に連続して形成され、上記のペットの尻尾を引き出すための孔が尻当て部に形成され、この尻当て部が孔から開かれて左右一対の分割片に形成されると共に、この分割片同士が接着自在に形成されているペット用の紙おむつであって、上記の孔が、その周縁にシャーリングが施されて収縮自在に形成され、上記の腹当て部の内側で且つペットの肛門寄りの股の位置に、糞の収納部が、肛門の側に口が開口されて袋状に形成され、この糞の収納部が、シート片を腹当て部の内面に口以外の周辺を固着することにより、有袋類の育児嚢状に形成され、シート片の口縁の中央部にシャーリングが横方向に一定の長さにわたって施され、口縁の中央部に生地が寄って口縁が収縮するよう形成されていることを特徴とするペット用の紙おむつ。

【請求項２】

請求項１に記載のペット用の紙おむつであって、尻当て部が、孔から開かれて左右一対の分割片に形成されるのに代え、ペットの尻の少なくとも一方の側を覆う状態に腹当て部に接続され、パンツ型に形成されていることを特徴とするペット用の紙

おむつ。

(2)【発明の詳細な説明】

(3)【技術分野】

【０００１】

本発明は、使い捨て式のペット用の紙おむつに関し、更に詳しくは犬等の尻尾を有する四足ペットに使用するペット用の紙おむつに関するものである。

(4)【背景技術】

【０００２】

従来この種の紙おむつとしては、例えばペットの尻尾を引き出すための孔状の開口部を背当部に形成し、この開口部に連通する切欠部を背当部に形成すると共に、切欠部で分離する両片を接着テープで接着可能に形成しているものがある（例えば特許文献１参照）。

【０００３】

ところでこの種の紙おむつは、尿が隙間から漏れ出ないよう尿を確実に捕集でき、また糞の後始末を簡単、迅速にでき、ペット自体も自分の排泄した糞でストレスが起きないよう形成されているのが望ましい。

しかるに従来品は、尻尾を引き出すための開口部と尻尾との間に生じる隙間から尿が漏れ出す、という問題点があった。

また従来品は、ペットが糞を排泄すると、糞がペットの尻や腹に付着し、ペットの体を汚すのを避けられなかった。そのため従来品を使用すると、飼い主は、ペットを洗う必要が生じ、後始末に手間暇がかかり、またペットにストレスを生じさせ易

い、という問題点があった。

【特許文献1】特開平7－236380号公報

【発明の開示】

(5)【発明が解決しようとする課題】

【0004】

本発明は、このような従来品の問題点に鑑み、提案されたものである。

従って本発明の解決しようとする技術的課題は、尻尾を引き出す孔からの尿漏れを確実に防止でき、また排泄した糞がペットの尻や腹を汚すことを防止し、糞の後始末にかかる手間暇を軽減でき、ペットがストレスを起こすことがないよう形成したペット用の紙おむつを提供することにある。

(6)【課題を解決するための手段】

【0005】

本発明は、上記の課題を解決するため、次のような技術的手段を採る。

即ち本発明は、四足ペット1の腹から後足2の股にあてがう腹当て部3と、尻にあてがう尻当て部4とが一体状に連続して形成され、上記のペット1の尻尾1aを引き出すための孔7が尻当て部4に形成され、この尻当て部4が孔7から開かれて左右一対の分割片4aに形成されると共に、この分割片4a同士が接着自在に形成されているペット用の紙おむつであって、上記の孔7が、その周縁にシャーリング8が施されて収縮自在に形成され、上記の腹当て部3の内側で且つペット1の肛門寄りの股の位置に、糞11の収納部10が、肛門の側に口10aが開口されて袋状に形成され、この糞11の収納部10は、シート片を腹当て部3の内面に口10a以

外の周辺を固着することにより、有袋類の育児嚢状に形成され、シート片の口縁の中央部にシャーリング１２が横方向に一定の長さにわたって施され、口縁の中央部に生地が寄って口縁が収縮するよう形成されていることを特徴とする（請求項１）。

【０００６】
本発明の場合、左右一対の分割片４ａは、通常、略同じ大きさに形成されるが、大きさを違えて形成されるのでも良い。またここで、接着自在に形成されている、とは、一対の分割片４ａ同士を、切り離し箇所を介してくっ付けたり離したりすることが自由にできる、ということを意味し、具体的には面状ファスナーや、剥離自在の粘着テープで切り離し箇所を接着することで実現される。

【０００７】
本発明は、上記のように形成されているため、これによると、収納部１０を、簡単に且つ嵩張ることなく形成でき、しかも腹当て部３の内面の吸水性樹脂の性能を低下することなく、この内面を活用して糞１１を確実に収納、捕集できる。本発明の場合、シート片は、通常、腹当て部３の内面と同じ吸水性樹脂材で形成されているのが好ましいが、これに限定されるものではない。またシート片の固着の手段としては、例えば縫合や、接着、溶着等がある。

【０００８】
また本発明の場合は、上記の通り、シート片の口縁の中央部にシャーリング１２が施されているから、これによると、シート片の口縁が腹当て部３の内面から離されて口１０ａが開かれ易くなり、糞１１を、収納、捕集し易くなる。

【０００９】
また本発明は、図４、図５に示されるように、尻当て部４が、孔７から開かれて左

右一対の分割片4aに形成されるのに代え、ペット1の尻の少なくとも一方の側を覆う状態に腹当て部3に接続され、パンツ型に形成されているのでも良い（請求項2）。

この本発明によれば、腹当て部3と尻当て部4が展開状態の場合に比べ、より簡単、迅速に装着でき、使い勝手が良くなるからである。ここで、少なくとも一方の側を覆う状態に、とは、ペット1の尻の両側を覆うように、腹当て部3から尻当て部4に連ねて形成されているのでも良い、ということを意味する。

(7)【発明の効果】
【0010】
本発明は、このように孔の周縁にシャーリングが施され、腹当て部の内側で且つペットの肛門寄りの股の位置に、糞の収納部が、肛門の側に口が開口されて袋状に形成されているものである。

従って本発明の場合は、孔の周縁のシャーリングが尻尾を締め付け、またペットが糞をすると、この糞を収納部が、収納、捕集する。

それ故これによれば、尻尾を引き出す孔と尻尾との隙間から尿が漏れ出ることを防止できる。また本発明は、排泄した糞でペットの尻や腹が汚れることを防止できるから、これによると糞の後始末にかかる手間暇を軽減でき、ペットのストレスを解消できる。

(8)【発明を実施するための最良の形態】
【0011】
以下、本発明を実施するための最良の形態を説明する。

本発明は、図1等に示されるように、犬等の四足ペット1の腹から後足2の股にあてがう腹当て部3と、尻にあてがう尻当て部4とが一体状に連続して形成されている。

【0012】
上記の腹当て部3は、長手方向に沿った両側縁の適宜位置に、雄型の面状ファスナー5aを有する舌片状の接着部5が、接着面を腹当て部3の内面と同じ側にして、2個づつ設けられている。この接着部5は、装着時に尻当て部4の外面の雌型の面状ファスナー6に付けられ、腹当て部3と尻当て部4とを連結するためのものである。この尻当て部4の雌型の面状ファスナー6は、ペット1の体形に合わせて雄型の面状ファスナー5aの止め位置を柔軟に変更できるよう、面積が大きく形成されている。

【0013】
また尻当て部4は、ペット1の尻尾1aを引き出すための孔7から開かれて左右一対の分割片4aに形成されている。孔7は、その周縁にシャーリング8が施され、ペット1の尻尾1aを軽く締め付けることができるよう収縮自在に形成されている。シャーリング8は、左右一対の分割片4aの、ペット1の股に対応する中央位置の縁に所定の長さにわたって形成され、孔7は分割片4aが尻尾1aを挟んで交差状に反されることにより、形作られる。左右一対の分割片4a同士は、この実施形態では略同じ大きさに形成され、接着自在に形成されている。具体的には、一方の分割片4aの切り離し箇所の側縁に、雄型面状ファスナー9aを有する舌片9が、一定の間隔をあけて2個設けられている。そしてこの舌片9が、各分割片4aの外面に設けられている上記の雌型の面状ファスナー6に付けられ、これにより分割片

4a同士が接着可能に形成されている。

【0014】

10は、糞11の収納部である。この収納部10は、上記の腹当て部3の内側で、且つペット1の肛門寄りの股の位置に、肛門の側に口10aが開口されて袋状に形成されている。この実施形態に係る収納部10は、腹当て部3の内面と同じ吸水性樹脂材のシート片を、腹当て部3の内面に口10a以外の周辺を縫合することにより、有袋類の育児嚢状に形成されている。また本発明品は、シート片の口縁の中央部に、シャーリング12が横方向に一定の長さにわたって施され、口縁の中央部に生地が寄って口縁が収縮するよう形成されている。

【0015】

次に本発明の使い方を説明する。

飼い主は、先ずペット1の腹に腹当て部3をあてがい、尻当て部4を尻から背中に回し、一対の分割片4aを尻尾1aを挟んで交差させ、一方の分割片4aの雄型面状ファスナー9aを、他方の分割片4aの雌型の面状ファスナー6に付け、尻当て部4でペット1の尻を覆う。次に腹当て部3の舌片状の接着部5を、尻当て部4の雌型の面状ファスナー6にくっ付ける。これにより本発明品が、ペット1に装着される（図1A、図2参照）。

【0016】

而して本発明品の場合、尿は、主として腹当て部3の内面の吸水性樹脂により捕集される。そして本発明では、孔7と尻尾1aとの隙間が孔7の周縁のシャーリング8により塞がるため、この箇所から尿が漏れ出すことがない。また排泄された糞11は、収納部10に落下して収納、捕集される。従って本発明によれば、ペット1

の尻などが糞１１で汚れることがない。本発明品をペット１から外すときは、上記の装着時と逆の手順で各接着箇所を剥がして行う。

【００１７】
以上の処において、本発明は、図４、図５に示されるように、パンツ型に形成されているのでも良い。図４に示される本発明品は、腹当て部３と尻当て部４が股の部分と、両側（図４Ａ、Ｂにおいて、左右の部分）とが、最初から接続されてパンツ型に形成されている。尻当て部４は、上例と同様、孔７から開かれている。ペット１の股に対応する尻当て部４の中央位置にシャーリング８が施されて収縮可能に形成されている。そして上例と同様、この位置が、尻尾１ａ（図５等参照）を引き出す孔７として機能するよう形成されている。

【００１８】
尻当て部４の一方の側縁には、上例と同様、雄型面状ファスナー９ａを有する舌片９が、この実施形態では一定の間隔をあけて３個設けられている。また尻当て部４の外面には、雌型の面状ファスナー６が設けられている。またペット１の股に対応する部分の両側に、ペット１の後足２を通すための開口部１３が形成されている。この開口部１３の周縁は、尿等の漏れを防止できるよう、シャーリング１４が施され、後足２を軽く締め付けることができるよう収縮可能に形成されている。また腹当て部３と尻当て部４の、図面上において上縁にあたる開口縁１５は、ゴム糸等で伸縮するよう形成され、装着時（図５参照）に、ペット１の腹回りを軽く締めて脱げ落ちないよう形成されている。その他の構成は、上例と同様である。

【００１９】
従ってこの実施形態の本発明品の場合は、開口部１３にペット１の後足２を入れ、

尻尾１ａを孔７の位置に配置して雄型面状ファスナー９ａを、尻当て部４の雌型の面状ファスナー６にくっ付けて接着することにより、本発明品を、簡単、迅速に、ペット１にはかせて装着できる。

【００２０】
なお、図６に示される本発明品は、同図Ａにおいて、尻当て部４の左側が、腹当て部３から切り離されている。この分割された尻当て部４は、雄型面状ファスナー９ａを有する舌片９が、雌型の面状ファスナー６にくっ付けられ、接着部５の雄型の面状ファスナー５ａが雌型の面状ファスナー６にくっ付けられることにより、ペット１に装着可能に形成されている。またペット１の後足２を入れる開口部１３は、図６Ａにおいて、右側は完全な貫通孔状に形成されているのに対し、左側は、尻当て部４の左側縁と、腹当て部３の左側の側縁との間の、シャーリング１６を施した箇所で形作るよう形成されている。この本発明品の場合も、図６において、上縁にあたる開口縁１５は、ゴム糸等で伸縮するよう形成され、装着時（図７参照）に、ペット１の腹回りを軽く締めて脱げ落ちないよう形成されている。その他の構成は、上例と同様である。

【００２１】
従ってこの実施形態の本発明品の場合は、図６Ａにおいて、右側の開口部１３にペット１の一方の後足２を入れ、尻尾１ａをシャーリング８の位置に配置して尻当て部４を矢示のようにかえしてペット１の尻を覆う。そして飼い主は、舌片９の雄型面状ファスナー９ａを、雌型の面状ファスナー６にくっ付け、次にペット１の他方の後足２をシャーリング１６が施された箇所に配置し、腹当て部３の舌片状の接着部５を、尻当て部４の雌型の面状ファスナー６にくっ付けて接着する。これにより、

図7に示されるように、本発明品がペット1に装着される。この本発明の場合も、上例と同様、ペット1にはかせて装着するため、ペット1が動いても本発明品を楽に装着できる。

⑼【図面の簡単な説明】

【0022】

【図1】本発明の紙おむつの好適な一実施形態を示し、Aは使用状態時の要部側面図、Bは内面側から見た斜視図である。

【図2】同上紙おむつの使用状態時の側面図である。

【図3】同上紙おむつの内面側から見た平面図である。

【図4】同上紙おむつの他の実施形態を示し、Aは使用例を説明するための平面図、Bは尻当て部を閉じたときの底面図である。

【図5】図4に係る本発明品の装着時の要部側面図である。

【図6】同上紙おむつの更に他の実施形態を示し、Aは使用例を説明するための平面図、Bは尻当て部を閉じたときの底面図である。

【図7】図6に係る本発明品の装着時の要部側面図である。

⑽【符号の説明】

【0023】

1　ペット

1a　尻尾

2　後足

3　腹当て部

4　尻当て部

4a　分割片

7　孔

8　孔の周縁のシャーリング

10　収納部

10a　口

11　糞

12　収納部の口縁のシャーリング

【図1】

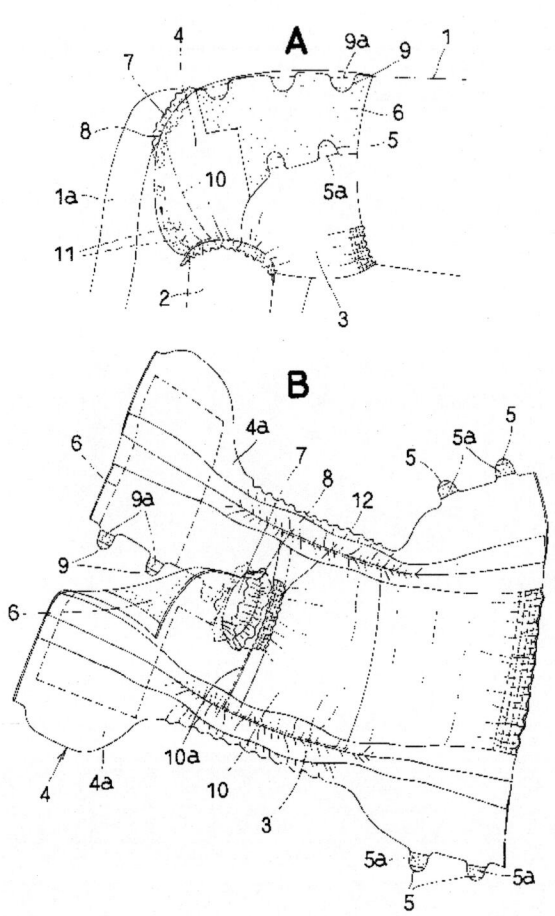

【図2】

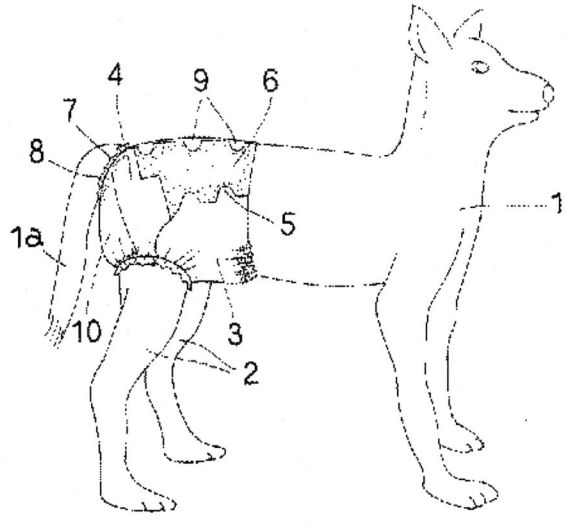

【図3】

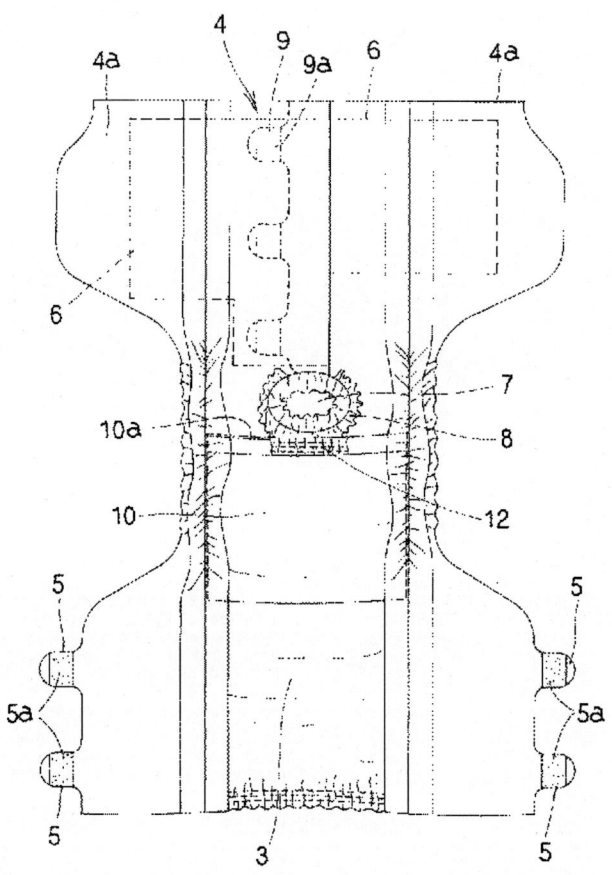

【図4】

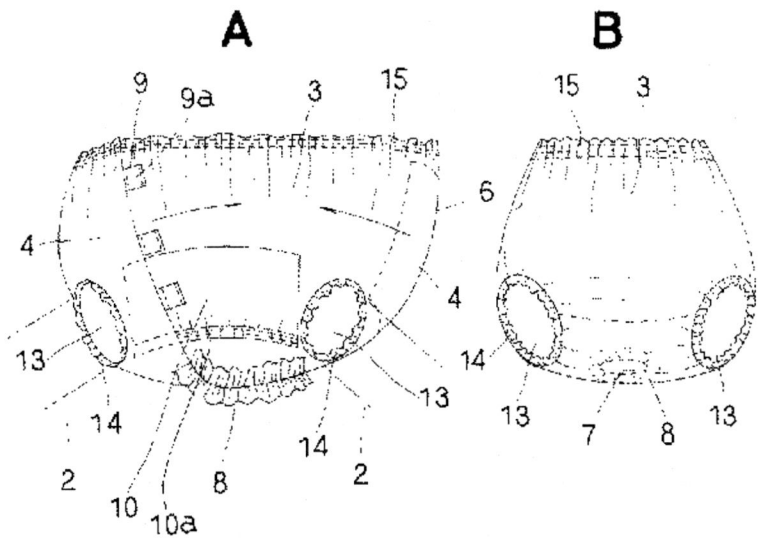

【図5】

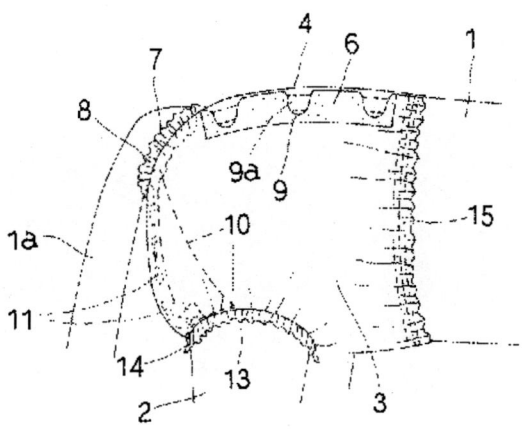

【図6】

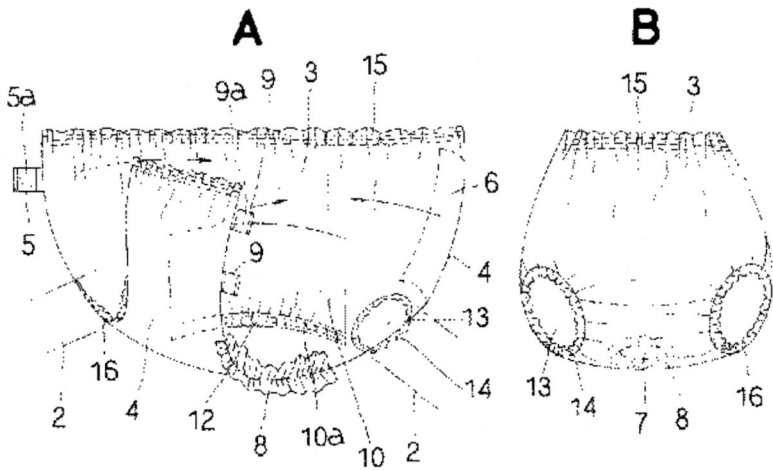

【図7】

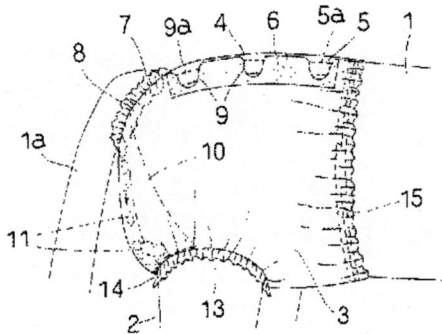

DETAILED DESCRIPTION

[Detailed Description of the Invention]

[Field of the Invention]

[0001]

The present invention relates to the disposable diaper for pets used for the 4-pair-of-shoes pet which has a tail of a dog etc. in more detail about the disposable diaper for the pets of a disposable type.

[Background of the Invention]

[0002]

Conventionally, as this kind of a disposable diaper, the opening of the hole shape for pulling out a pet's tail, for example is formed in a back rest, the notch which communicates this opening is formed in a back rest, and there are some which form both the pieces separated by a notch with adhesive tape so that adhesion is possible (for example, see Patent Document 1).

[0003]

By the way, as for this kind of disposable diaper, it is desirable to be formed so that stress may not occur with the stools in which could catch urine reliably so that urine might not be leaked and discharged for a gap, and the rearrangement of stools could be made simply and quick, and he excreted the pet itself.

However, elegance had conventionally the problem that urine begins to leak from the gap produced between the opening for pulling out a tail, and a tail. Conventionally, stools adhered to the pet's hips and belly and the elegance was not able to avoid soiling a pet's body, when the pet excreted stools. Therefore, when elegance was used conventionally, the owner will need to wash a pet, and the rearrangement took time and effort, and there was a problem of being easy to make a pet producing stress.

[Patent document 1] JP,H7-236380,A

[Description of the Invention]

[Problem to be solved by the invention]

[0004]

The present invention is proposed in view of the problem of elegance such conventionally.

Therefore, technical problem which is going to solve the present invention, The stools which could prevent reliably the urine leakage from the hole which pulls out a tail, and were excreted are prevented from soiling a pet's hips and belly, the time and effort concerning a rearrangement of stools can be reduced, and it is in providing the disposable diaper for pets formed so that a pet might not cause stress.

[Means for solving problem]

[0005]

The present invention takes the following technical means in order to solve the

above-mentioned problem.

Namely, the dinner-pad part 3 which assigns the present invention to the crotch of the hind legs 2 from the belly of the 4-pair-of-shoes pet 1, The hips reliance part 4 assigned to the hips is continuously formed in integral form, the hole 7 for pulling out the above-mentioned pet's 1 tail 1a is formed in the hips reliance part 4, and this hips reliance part 4 is opened from the hole 7, and is formed in the pair of right and left split piece 4a, and. These split piece 4a is the disposable diapers for pets currently formed enabling free adhesion, Shirring 8 is given to the periphery and the above-mentioned hole 7 is formed in it, enabling free contraction, In the position of the crotch of anus slippage of the pet 1, are the inside of the above-mentioned dinner-pad part 3, and the opening of the mouth 10a is carried out, it is formed in the anus side by the storage part 10 of the stools 11 in bag shape, and the storage part 10 of these stools 11, By fixing peripheries other than mouth 10a to the inner surface of the dinner-pad part 3, a sheet piece, It is formed in the shape of [of Marsupialia] a marsupium, and it is formed so that shirring 12 may be given to the central part of the peristome of a sheet piece covering length fixed in a transverse direction, cloth may visit the central part of the peristome and the peristome may contract (Claim 1).

[0006]

the case of the present invention -- the pair of right and left split piece

4a -- usually -- abbreviated -- a size is changed and formed although formed in the

same size -- you may have . Here, it realizes by meaning that it can perform freely carrying out beam **** with shoes of the pair of split piece 4a to being formed enabling free adhesion via a separation part, separating specifically with a plane-like fastener and the adhesive tape which can be exfoliated, and adhering a part.

[0007]

According to this, since it is formed as mentioned above, the present invention can form the storage part 10, without being simply bulky, without moreover being deteriorated about the performance of the hydrophilic resin of the inner surface of the dinner-pad part 3, utilizes this inner surface, and can store and catch the stools 11 reliably. As for a sheet piece, in the case of the present invention, it is preferable to usually be formed with the same water-absorbing resin material as the inner surface of the dinner-pad part 3, but it is not limited to this. As a means of fixation of a sheet piece, there are a suture, adhesion, joining, etc., for example.

[0008]

According to this, since shirring 12 is given to the central part of the peristome of a sheet piece as above-mentioned, the peristome of a sheet piece is separated from the inner surface of the dinner-pad part 3, and the mouth 10a becomes is easy to be opened, and the stools 11 are stored and it becomes easy to catch in the case of the present invention.

[0009]

It may replace with the hips reliance part 4 being opened from the hole 7, and

being formed in the pair of right and left split piece 4a, the at least 1 side side of the pet's 1 hips may be connected to a wrap state at the dinner-pad part 3, and the present invention may also be formed in the trousers type, as shown in Fig.4 and Fig.5 (Claim 2).

According to this present invention, it is because it can equip more simply and promptly and usability becomes good compared with the case where the dinner-pad part 3 and the hips reliance part 4 are expanded states. here -- the at least 1 side side -- a wrap state -- the both sides of the pet's 1 hips -- a wrap -- it is put in a row and formed in the hips reliance part 4 from the dinner-pad part 3 like -- **** -- what is said is meant.

[Effect of the Invention]

[0010]

as for the present invention, shirring is given to the periphery of a hole in this way -- the inside of a dinner-pad part -- and the opening of the mouth is carried out to the anus side, and the storage part of stools is formed in the position of the crotch of anus slippage of a pet in bag shape. Therefore, in the case of the present invention, if the shirring of the periphery of a hole binds a tail tight and a pet does stools, a storage part will store these stools and they will be caught.

So, according to this, it can prevent urine leaking and coming out from the gap

between the holes and tails which pull out a tail. Since the present invention can

prevent a pet's hips and belly from becoming dirty with the excreted stools, according to this, it can reduce the time and effort concerning a rearrangement of stools, and can cancel a pet's stress.

[Best Mode of Carrying Out the Invention]

[0011]

Hereafter, the best form for carrying out the present invention is described. As the present invention is shown in Fig. 1 etc., the dinner-pad part 3 assigned to the crotch of the hind legs 2 from the belly of the 4-pair-of-shoes pets 1, such as a dog, and the hips reliance part 4 assigned to the hips are continuously formed in integral form.

[0012]

The tongue-shaped jointing 5 of the edges on both sides in alignment with a longitudinal direction for which it has the plane-like fastener 5a of a male in a position suitably makes an adhesion surface the same side as the inner surface of the dinner-pad part 3, and the two above-mentioned dinner-pad parts 3 are provided at a time. This jointing 5 is for being attached to the plane-like fastener 6 of the female die of the outer surface of the hips

reliance part 4 at the time of wearing, and connecting the dinner-pad part 3 and the hips reliance part 4. Area is largely formed so that the plane-like fastener 6 of the female die of this hips reliance part 4 can change flexibly the stop position of the plane-like fastener 5a
of a male in accordance with the pet's 1 bodily shape.

[0013]

The hips reliance part 4 is opened from the hole 7 for pulling out the pet's 1 tail 1a, and is formed in the pair of right and left split piece 4a. Shirring 8 is given to the periphery, and the hole 7 is formed, enabling free contraction so that the pet's 1 tail 1a can be bound tight lightly. The shirring 8 is formed in the edge of the middle position corresponding to the pet's 1 crotch of the pair of right and left split piece 4a covering predetermined length, and the hole 7 is formed when cross form is contrary to the split piece 4a on both sides of the tail 1a. pair of right and left split piece 4a -- this embodiment -- abbreviated -- it is formed in the same size and formed, enabling free adhesion. Specifically, the tongue-shaped piece 9 which has the male plane-like fastener 9a opens a fixed interval in the side edge of the separation part of one split piece 4a, and is provided two pieces. And this tongue-shaped piece 9 is attached to the plane-like fastener 6 of the above-mentioned female die currently provided by the outer surface of each split piece 4a, and thereby, split piece 4a is formed so that adhesion is possible.

[0014]

10 is a storage part of the stools 11. This storage part 10 is the inside of the above-mentioned dinner-pad part 3, and the opening of the mouth 10a is carried out to the anus side, and it is formed in the position of the crotch of anus slippage of the pet 1 in bag shape. The storage part 10 concerning this embodiment is

formed in the shape of [of Marsupialia] a marsupium by suturing peripheries other than mouth 10a to the inner surface of the dinner-pad part 3 in the sheet piece of the same water-absorbing resin material as the inner surface of the dinner-pad part 3. Shirring 12 is given to a transverse direction covering fixed length, and the present invention article is formed in the central part of the peristome of a sheet piece so that cloth may visit the central part of the peristome and the peristome may contract.

[0015]

Next, how to use the present invention is described.

Assign the dinner-pad part 3 first to the pet's 1 belly, turn the hips reliance part 4 to the back from the hips, the pair of split piece 4a is made to cross on both sides of the tail 1a, the male plane-like fastener 9a of one split piece 4a is attached to the plane-like fastener 6 of the female die of the split piece 4a of another side, and an owner is a wrap about the pet's 1 hips at the hips reliance part 4. Next, about the tongue-shaped jointing 5 of the dinner-pad part 3, it is shoes attachment ** to the plane-like fastener 6 of

the female die of the hips reliance part 4. Thereby, the pet 1 is equipped with a present invention article (refer to Fig. 1 A and Fig. 2).

[0016]

It ** and, in the case of a present invention article, uptake of the urine is carried out mainly with the hydrophilic resin of the inner surface of the dinner-pad part 3.

And in the present invention, since the gap between the hole 7 and the tail 1a is

closed by the shirring 8 of the periphery of the hole 7, urine does not begin to leak from this part. The excreted stools 11 fall to the storage part 10, and uptake is stored and carried out. Therefore, according to the present invention, the pet's 1 hips, etc. do not become dirty with the stools 11. When removing a present invention article from the pet 1, it carries out by removing each adhesion part in a procedure contrary to the time of the above-mentioned wearing.

[0017]

In the above place, as shown in Fig. 4 and Fig. 5, what is formed in the trousers type may be sufficient as the present invention. The portion of a crotch and both sides (in Fig. 4 A and B, it is a right and left portion) are connected from the beginning, and, as for the present invention article shown in Fig. 4, the dinner-pad part 3 and the hips reliance part 4 are formed in the trousers type. The hips reliance part 4 is opened from the hole 7 like the upper example.

Shirring 8 is given to the middle position of the hips reliance part 4 corresponding to the pet's 1 crotch, and it is formed in it so that contraction is possible. And this position as well as an upper example is formed so that it may function as the hole 7 which pulls out the tail 1a ([Refer to / etc / Fig.5]).

[0018]

Like the upper example, by this embodiment, the tongue-shaped piece 9 which has the male plane-like fastener 9a opens a fixed interval in one side edge of the

hips reliance part 4, and is provided three pieces. The plane-like fastener 6 of the

female die is provided by the outer surface of the hips reliance part 4. The opening 13 for letting the pet's 1 hind legs 2 pass is formed in the both sides of the portion corresponding to the pet's 1 crotch. Shirring 14 is given, and the periphery of this opening 13 is formed so that contraction is possible, so that the hind legs 2 can be bound tight lightly, so that urinary leakage can be prevented. The opening edge 15 which hits on [of the dinner-pad part 3 and the hips reliance part 4] Drawings in an upper limb is formed so that it may expand and contract with a rubber thread etc., and at the time of wearing (refer to Fig.5), it is formed so that the circumference of the pet's 1 belly can come off in total lightly and it may not fall. Other composition is the same as that of an upper example.

[0019]

In the case of the present invention article of this embodiment, therefore, by putting the pet's 1 hind legs 2 into the opening 13, arranging the tail 1a in the position of the hole 7, and carrying out shoes attachment ****** of the male plane-like fastener 9a at the plane-like fastener 6 of the female die of the hips reliance part 4, Simply and promptly, a present invention article can be lent to the pet 1 and he can be equipped with it.

[0020]

In the Drawing A, as for the present invention article shown in Fig. 6, the left-hand side of the hips reliance part 4 is separated from the dinner-pad part 3. Of eclipse

****** with shoes, the plane-like fastener 5a of the male of an eclipse with shoes and the jointing 5 is formed in the plane-like fastener 6 of a female die for the tongue-shaped piece 9 in which this divided hips reliance part 4 has the male plane-like fastener 9a at the plane-like fastener 6 of the female die so that wearing to the pet 1 is possible. To the opening 13 which puts in the pet's 1 hind legs 2 being formed in Fig. 6 A in the shape of [with perfect right-hand side] a through-hole, left-hand side is formed so that it may form in the part which gave shirring 16 between the left edge of the hips reliance part 4, and the side edge on the left-hand side of the dinner-pad part 3. The opening edge 15 where it hits an upper limb in Fig. 6 also in this present invention article is formed so that it may expand and contract with a rubber thread etc., and at the time of wearing (refer to Fig. 7), it is formed so that the circumference of the pet's 1 belly can come off in total lightly and it may not fall. Other composition is the same as that of an upper example.

[0021]

Therefore, in the case of the present invention article of this embodiment, in Fig. 6 A, one hind leg 2 of the pet 1 is put into the right-hand side opening 13, the tail 1a is arranged in the position of the shirring 8, the hips reliance part 4 is returned like an arrow, and it is a wrap about the pet's 1 hips. And an owner arranges the male plane-like fastener 9a of the tongue-shaped piece 9 in the part where the hind legs 2 of the pet's 1 another side were given to the shirring 16 after shoes

attachment at the plane-like fastener 6 of the female die, and does shoes attachment ****** of the tongue-shaped jointing 5 of the dinner-pad part 3 at the plane-like fastener 6 of the female die of the hips reliance part 4. Thereby, the pet 1 is equipped with a present invention article as shown in Fig. 7. Since it can lend the pet 1 also in this present invention and equips with it, even if the pet 1 moves, it can equip with a present invention article comfortably. [as well as an upper example]

[Brief Description of the Drawings]

[0022]

[Drawing 1] One preferable embodiment of the disposable diaper of the present invention is shown, and A is an essential part side view at the time of a busy condition, and the perspective view which looked at B from the inner surface side.

[Drawing 2] It is a side view at the time of the busy condition of a disposable diaper same as the above.

[Drawing 3] It is the plan view seen from the inner surface side of a disposable diaper same as the above.

[Drawing 4] A plan view for other embodiments of a disposable diaper same as the above to be shown, and for A describe the example of use and B are bottom views when a hips reliance part is closed.

[Drawing 5] It is an essential part side view at the time of wearing of the present invention article concerning Fig. 4.

[Drawing 6] A plan view for the yet another embodiment of a disposable diaper same as the above to be shown, and for A describe the example of use and B are bottom views when a hips reliance part is closed.

[Drawing 7] It is an essential part side view at the time of wearing of the present invention article concerning Fig. 6.

[Explanations of letters or numerals]

[0023]

1 Pet   1a Tail   2 Hind legs   3 Dinner-pad part

4 Hips reliance part   4a Split piece   7 Hole   Shirring of the periphery of eight holes   10 Storage part   10a Mouth

11 Stools   12 Shirring of the peristome of a storage part

ペット用の紙おむつ 腹当て部型＆パンツ型の使用方法

定価（本体1,000円＋税）

２０１２年（平成２４年）10月20日発行

No. HA-027

発行所　発明開発連合会®
東京都渋谷区渋谷 2-2-13
電話 03-3498-0751㈹
発行人　ましば寿一

Printed in Japan
著者　橋本みち子 ©

本書の一部または全部を無断で複写、複製、転載、データーファイル化することを禁じています。
It forbids a copy, a duplicate, reproduction, and forming a data file for some or all of this book without notice.